Continuous Delivery

Continuous Delivery

Continuous Delivery

Through Orchestrated Engineering
and Principles of DevOps

Prasanna Singaraju

Copyright © 2016 Prasanna Singaraju

All rights reserved.

ISBN-13: 978-1537368832
ISBN-10: 1537368834
Printed by CreateSpace, An Amazon.com Company

"Loved it! Well written, Easy read, Thought provoking

Seems like my past and what I did here in VZ. Nicely explained the DevOps concepts and continuously delivery through 'The Chief'. Nicely brought up the concept of 'playbook', Like the Author's Note in the middle of the reading. Great Job. Very impressive!! I Won't be surprised if you become a Hollywood script writer and might end up getting a show in Silicon Valley"

- Sanjiv Gulshan, Director, Verizon

"This book depicts an important crossroad for development teams...stay with comfortable, yet outdated processes or radically improve delivery with 21st century concepts sure to increase collaboration, team productivity, transparency, and quality. Mr. Singaraju uses a novel-style approach that is engaging and relatable for anyone that has worked around technology delivery. I enjoyed the story very much. It is definitely worth the read…"

- James Frost, VP, Service Delivery, MCX

" Mr. Singaraju narrates the story in a simple novel like manner, while still getting the seriousness of Continuous Delivery across. A Compelling story line, Technology books with such a plot are rare, this one is a must read for everyone in Technology"

- Jaime Caro, Cisco Systems

"Simple, easy to read and the perfect way of telling a technology story. I could relate to the characters as I am sure many of you reading the book would. Continuous Delivery simplified"

- PV, Enterprise Architect

"Vow! Loved it! The first technology book I couldn't put down after I picked it up"
- Anonymous, Java Developer

" Very Good read, I like the way minute details are covered"
- Shreshta, IT Consultant

" Excellent!! I enjoyed reading the book. Good start with the opening scenario (a scenario that most people can relate to), it is what every organization, delivery team, leadership team goes through during a release phase. Great transition into the need for continuous delivery. Good emphasis on need for quality, and engineering principles that's really required for quality delivery, and not to forget 'metrics'. Content was very well balanced"
- Anonymous

"It was a different take on an IT topic, which I enjoyed. The story telling approach helped me relate to the situation and question my role to effect change in the organization. Was very compelling story."
- Micheal Weihl, IT Director

Continuous Delivery

Continuous Delivery

iDedicate

To My Dad, My best friend. I Miss you!

iAcknowledge

I am a little old school when it comes to writing; a mechanical pencil and eraser on paper is my style. This book wouldn't have been completed without my digitization crew Sanjay, Rashi and Purnima. Thank you folks for patiently reading through my scribbles.

My Mom and My entire Family for their patience and support always, Love you All.

My extended family "Madness" – You know why, especially Shreshta.

Sanjay Jupudi, My Partner, Mentor, Friend, Brother, Thanks for putting up with my eccentricities.

Rashi being a springboard and the friend to bounce my ideas, and being right by me to make sure I stayed on task!

Purnima, my wife, philosopher, my better half and most importantly, my best friend.

Vito, for sitting through the long nights with me, even though he had no clue what I was up to.

And All those that did not believe in me, Thank you, it made me stronger and a better person!

Continuous Delivery

Foreword
By
Dwight D. Moore

https://www.linkedin.com/in/dwightdmoore

In today's environment, IT departments are besieged with various principles, disjoint methodologies, and technical processes. Examples include DevOps, Agile, Continuous Delivery (CD), Continuous Integration (CI), Test Driven Development (TDD)/Acceptance Test Driven Development (ATDD), Behavior Driven Development (BDD), etc. An alphabet soup that is difficult for many to reconcile. The challenge for many departments is making sense of all the various techniques, while maximizing speed and collaboration.

Mr. Singaraju creates a vision through story-telling for quality engineering that assimilates various approaches. While never simple, his vision approaches the problem from an outcome based solution. It seeks to maximize collaboration, and leverage technology. It enables technology teams to view the broader picture to evolve towards greater effectiveness.

Directionally correct, and supplemented with high-level diagrams, this is a story that has the potential to help many solution teams improve.

*   *   *

Continuous Delivery

Continuous Delivery

Introduction

Continuous Delivery is a Novel set in an Enterprise IT backdrop. This is a work of fiction involving the key stakeholders in a large retail players' IT organization. For those of you in IT who read this book can relate to one of the characters in this novel or would know someone that would.

The story revolves around these key characters,

The Chief – Head of Technology
Dave – Director of Development
Tom – Director of IT OPS
Mark – Director of Quality
Phil – Senior Manager, Release Management
Brian – Business Guy
Pam – Director of the PMO

Yes, all the core roles involved in any IT

organization.

The intention of this book is to convey the importance of Continuous Delivery and the challenges encountered during the journey, in a novel like light hearted vein.

I hope you enjoy reading the book as much as I enjoyed telling the story.

Happy Reading!

- Prasanna Singaraju

PROLOGUE

It was a little after 9 PM on Thursday night. Dave's iPhone was in silent mode. The vibration made a buzzing sound and indicated that there were two notifications pending. The lock screen showed 2 messages. The first one from "Wifey" read, "Should I order pizza? I assume dinner is off ☺". The second one did not end with the smiley emoticon. It read, "Can I have an update??" It was from a number Dave had stored in his contact list as "The Chief".

Dave is the Senior Director of Application Development at a large retailer. His team built and supported a Web Application that brought in more than 60% of the firm's revenues. The past month has been grueling for Dave, his team and in fact the entire

IT organization. The Marketing team came back with analytics and quarterly report that showed an alarming dip in revenues two months running. The industry analysts reported how their competition had a record growth during the same time, thanks to a better usability and richer user experience that their modern design offered.

The directive from business teams and the chief was to make things right. Which essentially meant redesign, add a ton of new features and what the IT team felt like – A forklift upgrade! It was an important release in the firms history, probably the biggest they had since their launch several years ago. The pressure was obvious, thankfully no punches were thrown but the love this IT team once shared was evidently fading.

Today was a big day for Dave and the entire team there. It was supposed to be "The" big release that everyone was waiting in anticipation of. Well, "Was" because things weren't going right, not at that moment. To be fair to them, everything was awesome even last night at dress rehearsal on pre-prod. Everyone was pumped up this morning as they moved the builds over. And then, they started the environment checklist.

The environment checklist "passed with one anomaly". The database version in production was two versions behind all lower environments, pre-prod included. As part of the revamp planning, the Development team had determined the need to upgrade their database. They had sent an FYI to all teams, somehow the Release and Ops teams hadn't gotten the memo. It wasn't a big surprise considering the way this enterprise operated. They loved their silos to death. There were three major camps as Dave referred to them as. The Development Camp was Development teams and an unlikely ally – The Business, The Testing Camp had Program Management as an ally and the Release and Operations teams were one camp. Each camp was fiercely protective of their territory and information within the camp remained airtight leading to silos. They don't ask for information and they don't share theirs was their argument.

The conference room was getting stuffy though the thermostat on the wall showed 68°F. It was from the hostility amongst the groups it felt like. The bridge that had been set up for the evening was open for a while now. The spider phone on the conference room table showed the duration of the call to be 04: 41: 52 and kept ticking away. It wasn't supposed to be this long. The room echoed with a voice from the speakerphone "Anyone from the Development team on the line? Any updates?" Dave had stepped out a few minutes earlier to take a call from his Oracle account executive, Jeff. The call was a lifeline Dave had been hoping would work – A valiant effort to see if he could get a quick patch to support the older database version.

"I am Sorry Dave," said Jeff, "In fact, we had recommended Tom and team to upgrade several times".

Dave said, "I know Jeff, but the circumstances are different now, we need a solution. Perhaps we could upgrade.. "

Jeff knew exactly where Dave was going with it and interrupted Dave mid sentence with "That would be suicide, we've never even tried it before, plus we wouldn't recommend or support a hot upgrade". But Dave knew this before he was going to ask the question, no harm trying he said to himself. "Okay Jeff, got to go, Thanks" Dave said. "Sorry Dave, I couldn't help more …" Dave was gone by then.

As he walked back to the conference room, Kim the project manager cracked her lips in a sympathetic "I know how that feels" smile. She met Dave just outside the door and said "Tom's looking for you. We are getting awfully close to the 11PM cut-off, Decision time Dave". Dave nodded knowingly and signaled to Kim that he needed to take that, as his phone started to vibrate in his hand.

"This is Dave," he said. It was Kumar, his Architect and right hand man. Kumar said, "Dave – We successfully replicated the pre-prod database with production data. We can switch centers and use pre-prod as the live center while we address the database versioning in production" It was an option Dave agreed, a risky one at that, the impact though, he downplayed. Dave thanked Kumar and explained the strategy to everyone on the bridge. There were some protests on the phone, which were addressed, and Dave managed to convince them. The IT Operations team "requested" Dave who held the final "Go / No-Go" decision to provide a sign-off on the plan. The business team representative was just plain happy to get a release as promised by and to Marketing. Dave "signed-off" on the release as they switched centers.

Dave and everyone in the conference room let out a few expletives and exchanged high-fives. Dave pulled out his iPhone and went to the messages app. He opened the message from "The Chief" and typed "All is good, release success" and hit the send button. He watched the production servers along with Tom and team for a short while. He didn't see a reply from the chief. He Thanked the folks still out there, wished them a good night and walked over to the elevator bank and pushed the down button. The elevator music played on softly Chris Rae's Liquid Latin oblivious to the events of the night. Dave stepped out the elevator and walked to the car.

The dashboard clock showed the time to be 11:08 PM. Dave called his wife and apologized for change in plans and told her he was on his way. As he turned the first light, he picked the phone to check if he had any new messages. As he was about the enter the ramp, he navigated the car play system to last dialed, and was going to dial Kumar, but was interrupted with the car speakers playing his ringtone loudly. The name of the screen told him it was Tom – The IT and Ops Director. "This is Dave". "Dave it's Tom" the reply echoed in the car and had a strange urgency to it. "The alerts are going crazy, warnings, intermittent errors and now we have failures". Dave had barely crossed the first exit on the freeway and he went back into the right lane to exit the freeway, make a U-turn and head back into the office. He was still on the phone with Tom with instructions to rollback when he saw a new message pop-up on his phone screen. "What is going on here??" It was the chief.

Dave made it back to the conference room, ensured that the rollback was successful and typed on his phone "Roll back, release failed".

"7AM my Office" came the reply, almost instantly.

It was way past 2 in the morning by the time Dave finally slipped under the covers of his bed.

Well, that weekend they made amends and things were alright and both Blue and Green had the latest code, versions etc., but tomorrow (Well, today) morning at 7AM would be a turning point in the enterprise's IT organization.

* * *

1

After a night of tossing and turning, Dave finally stepped out of bed. The bedside clock showed 04:57. "Already" his wife said in a low throaty voice. "Got to meet the Chief," he said, half turned towards the restroom. He shaved and pulled out his favorite tweed sports coat, his wife had gifted him for their last anniversary. As he pulled out of the garage into the street, he couldn't help but think about the deadly turn of events last night. He pulled into the drive-through of his neighborhood Starbucks. "Good Morning, welcome to Starbucks, what can I get started for you today?" A friendly voice on the intercom asked. "Good Morning, I'd like a Quad Grande Non-fat Mocha please" He knew he would need those extra shots of espresso. "Anything else for you this morning?" "That would do it". "Window 1 please". He got his coffee and thanked the attendant. He took his first sip before he started driving away. He drove out of the Starbucks parking lot and turned left at the light and onto the ramp and joined the light traffic at that hour, towards the headquarters. The rush hour was still hours away as he drove at a steady 70 mph. The radio played his favorite Linkin Park tune. "In the end, it doesn't even matter" he screamed away. Dave smiled as his thoughts were interrupted and wondered what it was that really mattered.

He pulled into the parking lot and walked in the building entrance door. 06:32 AM the large digital clock behind the security desk displayed. Joey the night security at the desk who'd just wished him goodnight a few hours ago was surprised to see him so early. "Morning Dave! Early start?" he said with the same energy as he had said goodnight a few hours ago, obviously oblivious to the events from last night. Dave smiled and said "Sure Joey, you about wrapping up?" Joey pressed the elevator call button and mumbled "Good day". Dave said, " you too buddy" as the door closed. He walked onto the deserted floor and made his way to his office. He set his laptop on the docking station as it whined itself to life. He took a long sip of his Mocha, and wrote on the notepad his thoughts from the drive. 06:36 am, the screen saver showed.

The Chief sent an IM to Dave on Slack- their official instant messenger "Morning, You in the building?"

"Yes"-replied Dave.

"Lets meet in my office for the 7AM"

"On my way"

*   *   *

2

The Chief was held in high regard for two reasons apart from personality. His technology acumen and his amazing leadership skills. To many, his ability to stay calm and composed was out of grasp. The Chief had seen many enterprises through transformations the IT transformations of the 90's, Y2K, the more modern Agile and most recently digital transformation. Being an architect at heart helps him stay updated with the trends in technology and in fact, the Chief was the Architect behind this retail giant's e-commerce and technology platform.

Dave knocked softly on Chief's door. "Come on in Dave" he said. As Dave entered his office, The

Chief asked, "How's your little Quarterback doing". This in Dave's opinion stood out as a personality trait of the Chief that he appreciated most. Regardless of the gravity of the matter at hand, he was warm in his welcome and curious to know how Dave's family was. Dave talked as he settled down into the white armchair across from the Chief.

The tone of the conversation remained casual until the chief said, "So, Dave, Talk to me, tell me how we got here".

Dave gave a full account of the events over the past few weeks. The Chief listened intently with the occasional pause for the questions. He finally said, "Alright Dave, what next?" Dave began with the plan to make amends over the weekend, which the chief approved of.

Okay, it's about time, let us head that way. "We need some forklift changes here, Dave", said the Chief as they walked. They entered the 12th floor conference room called "The Star" that everyone referred to as "The Boardroom" where Tom, Pam, Mark, Brian and Phil were all waiting with their morning coffee. "Good Morning, Howdy Folks" said the chief. Everyone mumbled their good morning as Dave walked over to the windows to raise the blinds, the room was filled with the warm glow of the rising sun over the horizon.

The Chief assumed position at the head of the conference room table. He asked, "Any of you catch the preseason game last night?" and continued with "Thank God, we don't make software as brittle as our Quarterback's bones, he joked". They all knew he was an ardent supporter of the Quarterback in discussion. It felt awfully cold in the room that morning after the debacle last night, as all thoughts were on the potential slugfest, until the Chief used his humor to break the ice.

"Before I get started with anything else folks, I'll tell you it was one helluva effort last night, let no one take that away from you, too bad though, We didn't end up on the winning side of the release line. It was a great team effort and I am proud of it. Dave has me up to speed on your plan for this weekend and I approve of it. Thank you for the plan and volunteering to take care of business over the weekend, I appreciate it". Now, we are not here to talk about last night or the past release. It is in the books; well it will be this weekend anyways. We are here to talk about the future, the future of our organization. We can get there with changes, some forklift, others not so much. But we need to make that commitment today and here.

First things first, let me re emphasize that we are all players on the same team; more importantly one big family and we can't ignore that fact. We are all on the team with a role to play and we need to make sure we do it with utmost transparency. We are the platform for this company to stay "Numero Uno". We are losing to our competition for one and only one reason. We didn't make good use of our platform and technology. The margin isn't wide but before that wedge grows, we need to address things that have been conveniently ignored all along.

If we are to continue to stay ahead of competition, we need to go back to basics and look at the things that made us great in the first place. The Solution begins with us accepting that there is a problem. We need to bring more design thinking into our approach and the way we deliver applications to our customers.

At this point in the impassioned speech the Chief walked up to the white board and wrote

Mission Continuous Delivery

* * *

3

A few minutes before 8:00, The Chief's Administrative Assistant, Peggy walked in with breakfast and interrupted the meeting. The chief turned around and said, "As always, right on time Peggy, Thank You"

As they returned to their seats with their bagels and coffee The Chief resumed. "Now that we've got our mission statement established, let us now see how to get there". "Yes Tom" the chief said in response to Tom's "I have a question". "What exactly is Continuous Delivery? We already have DevOps". " Ah! I am glad you asked. "Let me explain," said The Chief.

The Chief had the knack of explaining the most complex things with such simplicity that his 90 year old grandmother could understand, when she weren't in her famous hallucinations. Today though, there was no reason for him to simplify. His audience was his Technology Leadership team. "Continuous Delivery" he said, "is creating an environment, an ability to produce applications or features within applications in short cycles - when I say produce, I mean developed, built, tested, and release ready, and have it all deployment ready. Automation is at the heart of it – automation across various different lifecycle phases. DevOps on the other hand, brings similar thinking and is heavily focused on process, recommended approach and the collaboration of seamless transition between phases".

"Any organization that has disparate teams for Development, Test, Release, Operations can relate to the pain we underwent last night. Application Development team's responsibility is to produce a high throughput of features, while Operations needs to ensure stability. QA and Release teams often times end up playing sides and end up in a tug of war with no significant role to play until something goes majorly wrong."

"Now, the goal of DevOps as a philosophy or Continuous Delivery as a process is to ensure the

Development teams learn to build applications that are stable and production ready while the Operations teams need to appreciate and learn agility and higher throughputs. While every stakeholder in the game, QA and Release included collaborates and maintains highest transparency. So, Tom is that Clear?"

"Yes" Said Tom.

"So, One thing is established now I hope – we don't have DevOps and for everyone's sakes Tom get rid of or repurpose your "DevOps Team". A DevOps team is fundamentally wrong and that term is an oxymoron," The Chief yanked at Tom.

"So, as the first step to our mission, the primary change is the culture. We need to build one that promoted Collaboration and transparency. Everything doesn't need to be big bang. If the developers can sit down to lunch everyday with QA and Operations and none of them walk away with black eyes, that is a huge start, in fact that like winning the football division!" Everyone burst out laughing at The Chief's humor.

"In all seriousness folks", the chief continued, "Let's make the most of the Agile culture we have built. Let us communicate and collaborate better starting from the planning /business teams to our production support team, educate the vendors, be it managed services or contractors on your team. We have now a new mantra One Team – One Dream. To accomplish our mission, we need to bring in this culture and it has to start now, top down from each one of us to this room and then it will trickle down to the grassroots. The core essence and success of Continuous Delivery lies in visibility. Visibility across and to all players on the team. I mean every member of the team from build, test, deploy, release, Operations and support. This enables a shorter feedback cycle and thus promotes agility to steer course as necessary and velocity for higher throughputs for everyone involved".

"The collaboration, communication and transparency is going to start with the way we interact obviously but more importantly lets start with a deployment pipeline going forward we will promote a culture of incremental changes and continuous improvement as a way forward. All of us in this room will be the steering committee for planning with the outcome making its way directly to the pipeline and our team members will all be abreast of the plan and the delivery system."

Everyone was upbeat by now and there were no questions, just relaxed expression on everyone's face, as they took a break.

Hey, there are no black eyes in the room, that's progress the chief thought as he smiled to himself.

* * *

4

During the break, Tom, Dave and Mark wondered how the wedge between them widened, and what had even caused it in the first place. They were buddies, who shared an after work beer, weekend barbeque, in fact their families did several camping trips together. They were all much more than colleagues. But now, they barely had time for any of the fun stuff. Happy hours after work were interrupted, they hadn't had a weekend off in well over 3 months, their siloed work style put a line between them, it turned to a wedge and now, it was realization time.

The three of them along with the chief were part of the original team when this retail giant had an early start with the e-commerce. They acknowledged each other with a smile, perhaps reading the others mind.

Mark said "the chief is absolutely right, I have a feeling Continuous Delivery is going to be the solution to the troubles we have" Tom and Dave nodded in agreement, Tom added, " I like it as well, I think the stuff about collaboration is spot on!" "This is going to be a long road, but I am looking forward to the journey" reckoned Dave.

As everyone filed back into the boardroom, the chief said, "Our success with Continuous Delivery would be heavily dependent on our ability to make adjustments to our process, methodology and the agility with which we are willing to adopt". There were nods of agreement in the room as The Chief continued in his low, passionate voice, his back to the room as he stared out the window at a flight that had just taken off. "What process will get us going on our mission and help sustain? Once we have a baseline process building on the methodology and nitty-gritty details would be easier".

I have a question, began Pam. "We just moved to Agile and created the whole thing on that, the team has finally come to terms with it, another process change…."

"Pam, that was two years ago" Brian interjected. "I am more concerned with the delays and what it might mean to our releases"

"Well it has been 18 months Brian, but my point is" Pam continued, "another change in process may not fit too well with the team, we are historically poor at adopting the new".

"You are spot on with that Pam but, I don't think we need disruptive changes, in my opinion what we need is to change the way things are done, in fact to your point Brian, if anything, our releases can be more frequent and of higher quality" Dave said.

" Yes and I don't think its necessarily adding more time into projects, it would be spreading the time we already take into multiple phases, and I believe one of the key things is we need more automation" said Mark.

All eyes turned to Tom; he had been quiet since being yanked by The Chief earlier. Tom began with "I think from an Operations and Support perspective, there are some things that we could change in process. We have just made some significant investments in Virtualization, Capacity, and Automation of environments – well to some degree anyways. My point being, we are geared up from a technology standpoint and ready to embrace changes as they come" He continued on, almost starting to sound more like the Tom they all knew, all passion and beginning to brag now, "We have uptimes close to 80% in lower environments and we have achieved five 9's in production"; He paused for the effect and continued, "Like I said, we are open to newer ways, but that would mean more budgets"

The Chief continued after patiently hearing everyone out, "Lets remind ourselves communication, collaboration and culture are the three things that are going to be the first step and of course like I said earlier, One Team thinking." "As far as process and methodology go," the chief continued, "I agree with Dave, We will not go for extreme changes. It doesn't make any sense, nor do we need to go that route. Mark, I like that mention about automation, let us hold on to that thought. What we really need is a tweak to the way we do certain things, and probably even do them well already. Tom, when do you typically get the infrastructure requirements when things change?"

"Well, it depends" began Tom. "Most times we have a call during the release week, usually on the Monday of that week, followed by an environment walk-through on release morning, but off late, the first we hear about new requirements is when we go over the checklist".

Dave added, "Well that, and we send out our planning team notes to Tom, Mark and Phil", to which Tom said, "The meeting doesn't always involve my team, the planning MOM doesn't always have a call for action". "Good point" said Dave, it had been something he was seeing more and more of.

"It is about time Communication meets collaboration at some point", The Chief said and added "and that exactly is the type of stuff we need to focus on and eliminate. The first tweak to process folks is the way we collaborate; All of us in this room will be the steering committee and the planning meetings will be held to include all the players here, with each of you taking responsibility to pass the plan to your teams, ensure sign-off by the following day after consulting your respective architects".

"More importantly", The Chief continued, "planning will no longer be a once a release event, especially if our mission is Continuous Delivery. We will switch this to be a weekly stand-up. We will call this Continuous planning". With no questions around the room, The Chief continued, "To be more specific, what I mean is, the monthly planning session is for the release at large, while the weekly sessions are to track progress, review feedback from the production". They all agreed to The Chief's line of thought. Brian had a question. "But Chief" he began, "We Shift-left I get it. But do we really need to have so many moving parts as part of the release, wouldn't it slow us down". The Chief smiled as he began to explain. "In fact, it is going to help us be better prepared" he said. "How many times have we had to react to production defects and leave out the original release items and shift focus? I am not saying the shift in focus would never happen again, but this approach will ensure that we are better equipped and this will even help us be proactive to understanding what our customers want, plus be on target to prioritize our release elements". Brian listened intently and nodded in agreement.

The Chief continued, "In essence a mere shift left wont cut it anymore, it wouldn't get us even close to where we intend to be. We need to shift-left but with focus on the right. We essentially need to adopt this new approach which we call Trend shift approach; we have to shift extreme left, to the planning phase. It is significantly more important to include all details needed, right when we start to think about stuff, to ensure we make every build deployment ready".

"We are not changing the principles of Agile, we already have Pam, and In fact we are building on it. We are going to make it better and more focused on what will work best for us. So, for continuous planning to be successful folks, we need to plan, adapt, and repeat."

"And remember folks, this is Step 1. If we are able to do this right, we are setting up a solid platform. It will get us where we need to be at.

In addition to this all areas of the engineering lifecycle must focus on 4 primary aspects:

- a) Collaboration
- b) Eliminate wait times
- c) Automate manual Tasks
- d) Deployment Ready Builds.

As a next steps folks, I want you all to regroup and get back next week with a detailed plan for each of the lifecycle phases."

As they left the boardroom, they did so with a new energy and excitement on what lay ahead. Each of them thinking of the roles and thanking the chief as they left.

The chief smiled to himself as he thought, That Went Well!

* * *

5

Friday night starting at 11, Tom's team had begun the upgrade process for the Blue and Green boxes, it had to be a staged process and if everything went well, they could be done in less than 3 hours for each box. This was the kind of stuff that made Tom nervous and as he liked to joke "Hot fixes and production upgrades are the reason I look ten years older, all this gray hair is from the stress".

Tom started the bridge for others to join while he was with his team in the conference room on the Technology floor. Tom announced himself and those in the room before he politely enquired to check who else were on the bridge. He gave the others a few minutes to join and when he heard from Oracle support, Mark's representative on the bridge, he kicked-off proceedings with a quick overview of the roles and the plan. Tom then gave his Data Center and server team the green light to go ahead.

The upgrade to Database was expected but one unexpected security patch to a couple of the web servers was identified as critical and announced on the call, which everyone agreed was the right thing to do. The upgrades were in preparation for the release slated for Saturday morning. Things went without incident and a little after 5AM, Tom announced, they were all set and live on both centers.

As he was getting ready to wrap things up for the day, Tom texted Dave as he stepped in the elevator, "All set, Your show buddy". Dave responded with a wink and asked him to get some sleep and that he would see him in a few hours.

Saturday morning, Dave told his wife he would take their Son to practice and that she could pick him up later in the afternoon. Dave's Son was training to be a Quarterback, and was the starting quarterback for his high school and also rated highly in their league division by his coach and other coaches in his league division. That morning as they were driving, He asked, "Dad what do you do at work?" Dave replied that he built software, so people could shop online.

"That should be easy right? Then why haven't you been home much this past week, Mom says you are working on a very hard project. Why is it so hard to sell something? Aren't your products good enough?"

Dave laughed at the innocence in the question. He tried telling him, software is complex buddy. "No!" His son protested, its not. Last week I built a robot on my Lego kit, its software that I had to write. It was easy!"

Dave laughed again as he pulled into the parking lot. "Football is complex" the little boy asserted, "but coach breaks down the plays and calls plays, it would be different for each down! We learn the playbook, follow the play call and boom! We Win! Why don't you memorize the playbook? Don't you have a coach who can tell you?" Dave promised him he would consider that.

As he was driving away, Dave smiled at his Son's question, suggestions and couldn't agree more, they needed a Playbook! Things went fairly smooth that afternoon with the release, barring a few performance issues, which everyone agreed, weren't show stoppers.

Dave held a quick team meeting with Kumar and the other members on his team. He told them it was a fine job by the team and how much he appreciated all the hard work. Kumar and Mark – Dave's Architect's asked for updates from the leadership meeting. Dave gave them a rundown, Kumar and Mark hung onto every word.

Kumar finally said, "Agile, DevOps and now CD? What is different now Dave? So, how different will this all be for us?"

Dave said, "We will work out a detailed plan next week fellas. I am going to set up the second half of Monday as a working session. Let us talk in more detail then, but for now, I would like for you to come prepared to discuss the current team structure and sprint plan."

Tom, Mark and Phil had similar debriefs with their teams. There was excitement and nervous nods all at the same time. There were changes coming they said, just not sure how intense. While the leaders assured them quietly that all this change was for the good.

The Chief was glad the release went without incident. He sent off a quick kudos to the team.

The rest of the weekend was quiet and everyone looked forward to Monday morning.

* * *

6

At their morning IT-Operations huddle-up, Tom was in a particularly cheery mood. He told his "Boys" as he liked calling them. "We are going to enjoy this boys" the excitement and energy in his buoyant tone obvious.

"We are going to be agile" he continued, "and we are going to automate the heck out of this whole process. This past weekend was rough and you boys have done us proud. With a few tweaks to the way we do things, we can get better than ourselves. I want us to collaborate better with teams across the board". He went on to explain the full details from the technology leadership meeting. The team showed a little excitement and a lot of apprehension, which was to be expected.

Then, there it was. The seemingly obvious question looming on everyone's mind popped from Matt. "So, what happens to DevOps and what happens to the DevOps tools we have spent so much time with; All the adoption and efforts on research seem to be a waste of time, if we are going to start a new thing called Continuous Delivery?"

Somehow these folks believed DevOps was a team and their organization had a DevOps Team and DevOps tools. Tom paused before answering, "Let me just say, we will repurpose as appropriate. We are going to see where those folks can add more value". Tom concluded the huddle up with "I'll reach out to you boys as I work on the detailed plan this week, but trust me boys, this will be fun!"

At the same time as Tom's meeting, Mark was having his morning stand-up. That morning, Mark wasn't as cheery as Tom was. He was getting a debrief from the Performance Test Manager on the performance issues they had encountered. "We did not have the environments and time Mark," She said. "Most of the issues were what we had highlighted last time, they chose to ignore us and did not fix them" added Phil, one of the leads on the team. "What sort of issues were ignored?" probed Mark.

"Well, for starters, we run load tests only from the client side, we are not equipped to handle the database, connections or the server side, neither from a skillset perspective nor the toolset needed" the test manager ventured to which Phil added "most issues from last night were due to connection times and Database slow downs". Mark dismissed the subject with, "we need to come up with a plan to handle this better".

"Let us move on to Automation," he said. The automation team started with the total number of scripts the team had developed and explained how much of effort was being spent on improving the framework and such. Mark being the practical guy that he was, asked pointed questions around test coverage, the script utilization, reduced manual effort resulting from automated testing. The automation team had no answer to any of those questions from Mark. There was awkward silence in the room, which was finally broken by the Test Lead representing the vendor team for manual testing. The manual test lead started with showing off several charts highlighting coverage, task completion and functionality covered. Mark was quick to point out that the functionality tested was overlapping with the automation scripts that were being built. "Yes, but" protested the test lead from the manual testing team, "We did not even have the information on the total available scripts and that they did not have confidence on those available for execution within the repository because most of them failed". This annoyed Mark further. He looked to the automation lead who pointed out that the failures were due to environment issues.

So, Mark thought, we have a ton invested in automation and performance testing; Automation scripts aren't being used as much as they should be and performance testing is not being performed with an intention to understand system behavior – Man this isn't good! He thanked everyone for the updates.

Mark then said he had some updates for them from the Technology Leadership meeting. It was expected, especially after the events from last week and the weekend. Everyone sat up in his or her chairs now, fully alert.

"We are going to go extreme left folks. We are the one team in the value chain that is best equipped to call the shots; The team that can enforce quality checkgates, drive quality upstream and the team that is most empowered to keep everyone else honest about Quality. We need to up our game and be proactive. Automation is going to be mainstream and the theme for us going forward. Continuous Testing is going to be the mantra for us and we will be the one team that shifts both right and left. How? Let me explain"

"Going forward, we will be involved in the lifecycle starting at the planning phase, work collaboratively with the development team in an approach such as Acceptance Test Driven Development (ATDD), build our automated test scripts as the development team completes their process, conduct performance and security tests within that build sequence, and execute all the tests once we have the build ready. Post that, we will essentially work with IT-Ops to identify the top business processes and provide them with scripts for functional and user experience monitoring which we would then monitor, analyze and report into the continuous planning phase. Simple enough, right?"

There was silence in the room, the kind that is scary. He waited a full minute before he heard the first response.

"Sounds great Mark, but we don't have the skillset for it" This was the theme of comments as he went around the room. Mark smiled, this was half the battle won. There wasn't resistance, but it was more of an Ok, but how? kind of reaction.

"Folks, I am not saying this can be done overnight, nor do I mean to imply that this is easy. Collaboration is the first step, equipping ourselves with the right toolset and skillset as a team is next and the final but most important piece of the puzzle is eliminating the manual activities – I get it that manual testing is essential, but let us look for areas we can eliminate those ridiculous wait times and keep manual activities to writing acceptance test cases and measuring the user experience" Mark could see everyone in deep thinking modes and some smiles in the room. He wrapped up the meeting and told them that the next step would be a working session with the leads to focus on specifics.

Phil owned the Release management. He was very young in the company and this past week was his first release with this team. He was yet to entrench himself completely. The week he got here he was amazed at the organized chaos and was actually surprised that things were getting done here. So, in a way this was kind of an ideal situation for Phil. He had been an advocate of automated deployment and release for a while now but had little to no traction. One of his key asks was to get a production like environment for pre-production and to further have better configuration management practices across this pristine environment. From his time in the group, he had constantly looked to bring in a culture where a deployable binary would be available in pre-production environments periodically, rather than have everyone geared towards one release and hope for the best. He was a strong proponent of small, incremental changes and releases that gear towards continuous improvement. He had requested for a meeting with technology leaders last week but that got moved due to the firefighting that everyone was busy with. He looked forward to working towards a better future and the Mission Continuous Delivery was the best place to start.

Dave canceled his morning stand-up that day and put a hold on all new development activities until further notice. He instead had a working session scheduled with his own team.

* * *

Author's Note

Hello there! Oh yes you, with the book in your hand. I am in fact talking to you. I hope you are enjoying reading the book as much as I enjoyed bringing it to you. To get to Continuous Delivery, some of the more essential changes to be made are around people – Collaboration, skillset, perfect hand-offs; Ecosystem – Top Down drive for Continuous Delivery, Adoption of technology, Elastic Environments, Stronger Pre-Prod and automation.

But the most important piece of puzzle for an enterprise aspiring for Continuous Delivery is a tweak to the Process – An Agile and Lean process that empowers teams and enables them to collaborate with mesmerizing harmony like an Orchestra is the key to succeed. It has to be a culture that promotes One Team – Uniform process. A mere shift left in process alone will not cut it any more.

In this work, the focus will be to touch on each aspect of the Continuous Delivery Value Chain and how an Enterprise like yours worked towards getting there.

*   *   *

7

As the last Tech team lead on the Invitee list responsible for the Search module entered "The Great Divide" conference room at 11:31 AM, Dave said, "alright, that is everyone," as he went towards the door and shut it.

"Alright fellas, good morning! I know this has been a short notice to drop everything and spend the rest of the day in a heads down working session, I say heads down meaning literally, no gadget of any kind, cellphones and apple watches included. He began by summarizing why they were all there.

He walked over to the white board and wrote what the Chief had written earlier in their discussion.

Continuous Delivery

He explained their mission statement and what it really meant.

He turned away from the white board and continued, "As part of this journey, there are going to be things we haven't done prior to now as part of development. There are areas where we would mend our ways, areas where it might seem like a distraction and I am open to debate if there is a better way of doing it than what I am suggesting here. We will take a Collaborative Development approach. We will break things down, meaning – requirements / user stories will be broken down further into smaller units, which basically would be features; the features would further be broken down into scenarios, that's not it, the scenarios will be broken down into steps. The steps essentially translate in development terms to methods or units of code developed by you".

He took a pause to let what he had just said settle in, and then asked "Questions?"

A hand had begun to rise, tentatively. Dave said, "Yes, Jason, Is a question forming or are you ready?"

Jason was one of the leads on the team who always had something to say. I wouldn't call him opinionated, grumpy but lets just say he wasn't the happiest guy on the team.

Jason began, "Dave, the time we have is barely enough to develop code, validate it, and get it checked in, forget the integration issues, part of the problem is we underestimate time needed. We underestimate complexities, and to top things off the automated tests fail and they make us analyze their defects which end up being scripting issues by QA team. Anyways, my point here being, with such a time crunch already, why are we introducing these new measures? Why break things down? We develop units at the end anyway right? Plus we already test our code and make sure it works on our machine before we check-in. I don't see how this is different and why this is even needed."

He sounded extremely angry if you didn't know the guy; but here, it was just an innocent question.

Dave smiled as he gathered his thoughts. He said, "Jason, I don't mean to offend anyone here, you guys do a great job and I appreciate it. How many times during the past release have you picked up a user story built something over 8-10 hours only to be told later in the week during the demo that a you missed half of what the BA meant for that user story? How many times have you ended up adding those you missed to future releases/backlogs? Any of it sound familiar??"

Jason grinned sheepishly and agreed that it was more than a few times, "But Dave" he said "those were due to deficiencies in the requirement had nothing to do with me or us"

"Hold on to that thought Jason", Dave said. "You always test your code —Yes, you do. But when you check-in your code and it breaks the build or doesn't play well with others, you end up spending more time fixing it". Actually we estimate very well but when implementing is where we spend a ton of time. On an average we spend 30 to 40% of time coding and the rest fixing issues.

Kumar laughed and said, "Dave is absolutely right Jason on both counts actually. I see potential benefit in this approach"

"But Dave" Kumar added, "Jason has a point, we don't have bandwidth to handle the part where we have to break requirements down, and analyze. It is too much documentation." Dave explained, "Guys, first this isn't changing the way you develop code and next this isn't documentation, it is a change in the way we approach code development"

Dave continued, "We will adopt a collaborative development effort. Remember what I said about collaboration earlier? We will work with QA folks. We will bring in extreme pair programming techniques, I know I just made that term up, but it is exactly what it is". "I like that term as well, but specifically, how do you mean?" asked Kumar.

Even before he said what he was about to say next, Dave knew this was going to be an item of debate and probably controversy even. "The testing team will join us in building a collaborative Development process and they will bring that additional bandwidth we need. They will set up automated unit tests for us and will set up the steps that need to be developed into methods or units", said Dave.

"Oh Come On! We all know they ain't no developers, they ain't no techie folks to do that kinda stuff", said Tony. "All they can do is click here, click there, and Pass test cases", said another voice. Dave was ready with his response as he saw this coming. "Well fellas, it is a skill and it requires some coding experience, I get it. I am going to work with Mark and team, once we go this route, we will get cross skilled resources and this is one way to make this whole approach successful"

"Guys remember we need to change the way we look at it. We can no longer depend on Testing as an activity at the end of the development cycle, we need to ensure that delivering high quality code is our responsibility".

Dave walked to the front of the room to the whiteboard. He started making a list below where he had earlier written **"Continuous Development"**.

1) Test Driven Development
2) Acceptance Test Criterion
3) Principles for early testing
4) Automated Unit Testing
5) Nightly Builds
6) Build Validation and Continuous Integration

As if on Cue, Mark walked in as Dave finished explaining the process tweaks to his team. "Welcome Mark, thanks for stopping by." Dave quickly walked Mark through the methodology and discussed the areas he needed them to work together on. Mark loved the idea.

As the plan was forming on how the process must be for this extremely collaborative development methodology, Kumar brought up a good point – "But Dave", he said " Test Driven Development I thought is a technique for developers to brainstorm on a particular piece of development he or she is about to undertake. Moreover, I think Acceptance Test Driven Development or Behavior Driven Development or any of the Double D's are nothing Development done right, how specifically will a QA team member be able to help ?"

To this Dave said, "let me explain" and he referenced (and improvised) from an article he had read years ago by Rob Myers, to explain how it was supposed to work.

"If your mission were to build a Sports car – not just any sports car but the ultimate driving machine, how would your agile team handle this. Your thought process or sequence might be what is the purpose of this car? What would customers want; Excitement, power, safety, comfort and you make your list thus.

For Test Driven Development (TDD); Acceptance Test Driven Development (ATDD) or Behavior Driven Development (BDD) the primary benefit is Collaborative Development for higher quality code. Collaboration between all stakeholders in the value chain. Let us look at the approaches in which you might do this to understand the difference.

So, if you were to take this up as TDD you would start off with creating some unit tests; whether these resemble the actual unit tests is debatable. But as a customer of BMW, let me take a shot at what the engineer might come up with.

a. When the piston reaches a height of X, the spark plug fires
b. When the brake pedal is pressed 75% of the

way, the extra bright LED tail lights would be activated
c. When braking the Anti lock braking System needs to be activated
d. It needs to have 400 hp so it could be x, y or z

Now, try walking the business team or the scrum master through these TDD steps. It would be extremely hard. Though it makes total sense to your team and you. Some details described are highly technical in nature, though great info it's better left under the hood, and they need not be conceptually accessible to customers or business users.

Now consider adding a different perspective to this. We start with the same thought process as before. However, lets take a more user or customer centric view to this. This would be close to what it might look like.

a. When I punch the accelerator, I am pushed back into a plush comfy seat and amazing acceleration
b. When I slow down the breaks, the car comes to a stop without skidding or flipping and drivers behind me don't run into my rear end
c. When I drive in rain, I shouldn't skid off the road, the experience must be consistent

So the statements are both intended to do the

same thing. Breaking down into smaller units what the larger goal might be. The second set of tests conveys the same message, is business friendly and also is the acceptance criterion for a great driving experience. They are clear, small and repeatable tests that anyone can perform and would expect to pass each time. This is still TDD but more acceptance criterion driven.

These small incremental tests become a part of your regression and baseline criterion as you move forward with building other features and functionalities. Now the previous list by engineer (TDD) are important from an architectural conversation, systems and subsystems. The focus is different. ATDD is both business friendly and team guiding while TDD ones are micro or unit tests that developers use.

The micro-tests or unit tests are what you would build as you develop the code using the Acceptance Test developed by the testing team. So, we need a combination of the two. Let us call it "our process" for now.

ATDD is business like providing guidance to the development team, while TDD is too technically focused not business or tests friendly. What we need is a combination of the two that can help both sides. We need something that can assist us with the ability to define acceptance tests, use those as guidelines to then develop."

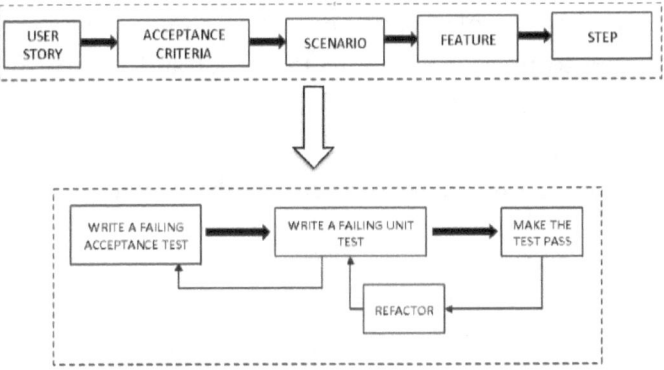

As Mark and Dave continued their discussion around the roles and parts that each member of the team would play several key discussion items emerged. It was exciting just to see them all in action tweaking the process as they went and making the big picture block diagram into more of a Continuous Development blueprint. Everyone who raised brows, hands and apprehensions earlier was now a contributor and more importantly a " believer".

They all agreed that Mark would work with the automation folks on his team and the QA vendor to build a more cross skilled team, that would essentially work more closely than ever before with the development team. It was going to be One Team!

Dave thanked Mark and so did the entire team in unison.

Dave and Mark shook hands and were visibly excited at the plan and the days to come. They both agreed to present jointly to the Chief and rest of the Technology Leadership.

Continuous Delivery piece 2 of the puzzle, check!

* * *

Author's Note

As they walked away from the meeting that day I wondered wasn't it always meant to be this way!! I see an engineering team in the making. Applications have to be engineered after all!!

*   *   *

8

Tom gathered his boys in "The Journey" conference room across the floor from where Dave had his meeting going. He began by saying, "The last week and weekend have been rough fellas, good work though, I couldn't have asked for more under the circumstances. One of the key things I take as lesson from here is, we should be more proactive. Mind you it doesn't take anything away from the effort and success from this past weekend. I intend for us to be the team making proactive recommendations. Little things starting from the Versions, hardware, Performance tuning measures, production feedback and much more"

The energy was high always with the boys; he never had issues with that. They were all high energy, young and a passionate bunch. "We are the team" he said, "that is the most visible, the most savvy with Information Technology and with the best ability to make an impact to everyone in the value chain, in fact I'll go ahead and say this, and perfectly equipped as well" He paused for the effect and resumed again.

"How? Let me explain".

"Let us start with some of the high-level things we own, control and oversee and the answer will appear very simple. We build, support and maintain all the environments, downstream, upstream, integration elements, connections, data center, tools, deployments, production support and more. Stakeholders, both internal and external depend on the infrastructure we provide. In a way, what our enterprise is to our customers is a direct reflection of how well we do.

So, We ought to be the team that collaborates best with everyone in the value chain. With internal stakeholders we can talk and get direct feedback, with clients we don't talk directly, but we have a powerful tool that can help us interpret what the customers are saying. Its called the log files. We already know what we are good at, Infrastructure, best in class environment uptimes and monitoring and product support."

"Now, it is time for us to learn from others along the value chain, learn what they are good at, Agility and Incremental and continuous improvement."

The boys were smiling by now. There was a side conversation brewing with Mahesh, one of the senior folks on the team itching to talk. He had a tendency to do that. Tom almost sensed what that conversation might be about. The boys must have picked up only the words he said, "learning from other teams" and the word "agility".

"What is it Mahesh?" Tom's voice thundered in the conference room. Mahesh, was a Virtualization expert and a hardcore infrastructure guy said, "Tom, what I was just telling here is that we brought down our turnaround time from 3 weeks to 1 week with virtualization and a ton of effort on all internal environments. I think while further optimization is possible there isn't more agility we can get".

"Oh, come on! Mahesh" Tom snapped. "I don't want to hear what has already been done. Think of eliminating that one week time down to a few hours"

Tom continued without waiting for a response, "There is so much we can do. Think about it boys, that one week wait time that Mahesh was referring to is a bottleneck right there. There are several of those. For instance, our glaringly obvious bottleneck, which then obviously becomes a bottleneck for all other teams, is the continuous availability of lower environments, specifically Development and QA. Now consider how long we take in standing up and provisioning new environments, what if we could make it on-demand; The completely out of sync Pre-Production environment, imagine the possibilities there, production analytics, rather the lack of them, which makes us reactive and so much more! Simple boys, let us make a list of things we can come to be a better operations team".

He continued, "let us keep all of those things in mind and make our primary focus to be around agility and automation". One of the biggest challenges for us has been the breakdown in communication, getting informed last minute, not being in the loop all along. We can now avoid all those, one of the tweaks we would make to the process that will positively impact everything else we do is, We will shift far left and be prepared sooner in the lifecycle".

One other thing that we need to move our focus towards is Release and deployment Automation, the addition of Phil will greatly boost this area, this is something that we will implement, and enable as a team. Let us leverage pre-production better and make it production-like at the earliest, specifically with respect to the controls we have on that environment. Let us explore containers and things like that will enable our development teams and others to build applications in production like environments".

Then the bummer for the team as he announced, "The DevOps Team is officially dissolved and we will repurpose all members as needed".

This was classic Tom, rumbling away. The team was trying to hang on to everything Tom was saying and luckily for them he spelled out the different work streams and nominated folks for each when he walked to the white board and wrote: "Team Goal: Release & Deployment Automation" and below are our individual focus areas:

1. Lower environment Stability and Management
2. Pre production Environment (Enhancements)
3. Toolset for Release and Deployment Automation
4. Leverage Cloud (In Lieu of lower environments)
5. Monitoring and Feedback analysis
6. Containers and other ways to eliminate surprises
7. Collaboration, Shift left with focus on the right
8. Configuration Management

Let us come up with a model for all this on how we

would achieve our goal. I'm going to meet with all with more detailed plan soon."

He finished off with, "Are you boys not tired of losing out on your happy hours?? I sure am! Let us win them back, one release at a time."

*　　*　　*

9

Since his meeting with Dave and the Development team, Mark operated with a new energy and purpose. He summoned his managers and leads to "The Longhorn" conference room. Mark had a few minutes to compose his thoughts before the team arrived. He made a mental note of how he planned to address this subject with the team.

Change is good, but if the case is not presented appropriately it might drive insecurity and cause a lot of unnecessary heartburn and distraction, he reasoned with himself. His team was good with change usually, but the recent team event where he announced the organizational change coming hadn't been the perfect setting. He knew that this shift in culture that the leadership was planning demanded a lot more than people simply adapting to the change. For his group, it meant a new brand image.

Mark knew it, even though no one really said it to his face, what others thought about the current Testing team he headed. The rest of the organization looked at Mark's group as a "Not so technical bunch that only followed instructions" In Mark's opinion this was never a good thing. His "go-to person" on the team had agreed on multiple occasions, but by themselves, they weren't able to change too much due to time crunch he reasoned.

"QA is often the red headed step child" Mark would say, "never enough time for it, even when QA did something right, people always have a way of politely ignoring it in lieu of other things being done"

Mark and three other folks from his organization that were FTEs had attended an Agile Transformation session and implementation from a QA perspective last quarter. When he came back and presented to the technology leadership, announcing he intended to do things differently, all he got in his opinion was a cold shoulder and a polite, we don't have the time to invest in and train you all. In fact, Mark was instrumental in hiring Phil, he had pushed for a Release and Deployment manager who could assist with automated build validation, the term he had heard was CI that truly felt like a super power they must have.

His joy knew no bounds when Dave had explained how critical Mark's team was to the success of Continuous Development. Extreme collaborative development, just that word still gave him Goosebumps. He was a man on a mission. On a mission to transform his group into Engineering driven, Engineering inspired Quality group that was going to up the game and standards for themselves and be a powerful ally to the rest of the organization they supported.

As his team filled the conference room, he welcomed them warmly and told them that this was a follow-up session, a session that would change the way they would approach and UNthink conventional testing and Quality Assurance.

Inspired by the meeting with Dave and team he called his Team Goal " Continuous Testing". " This my people is going to be our mantra going forward. We will actively be involved in the application development phase as the folks that enable and assist with Test Driven Development, a methodology that will bring higher quality code with every check-in, every integration and eventually every build. Not just that, our goal will be derived through a combination of Unit test automation, Early functional automation, and early non functional engineering which would include performance validation for each build, security tests for each build and cover accessibility and compatibility aspects. We will get as close to 100% in-sprint automation by taking up and working closely with the development team. In essence, we as a team will play at different levels early on in the lifecycle and again play a crucial part in the end game.

For that, we will need to think of some internal restructuring ourselves. I believe there would be four primary focus groups we would need to be.

a. SDET's as the TDD / BDD (in-sprint)
b. Automated Testing (in-sprint and regression)
c. Non functional (in-sprint and regression)
d. Support (Resource Management, Environments, Innovation)

"We will be a Quality Engineering team that is going to preach a new philosophy to everyone, Quality is a responsibility. Yes, RIP testing as an activity!"

Our goal will be to ensure we have a deployment ready build at any given point during the sprint. How? By doing a, b, c on the white board to perfection and providing a validated build to Phil and gang. Here is a representation of what it might look like, he pointed at the screen.

Continuous Delivery

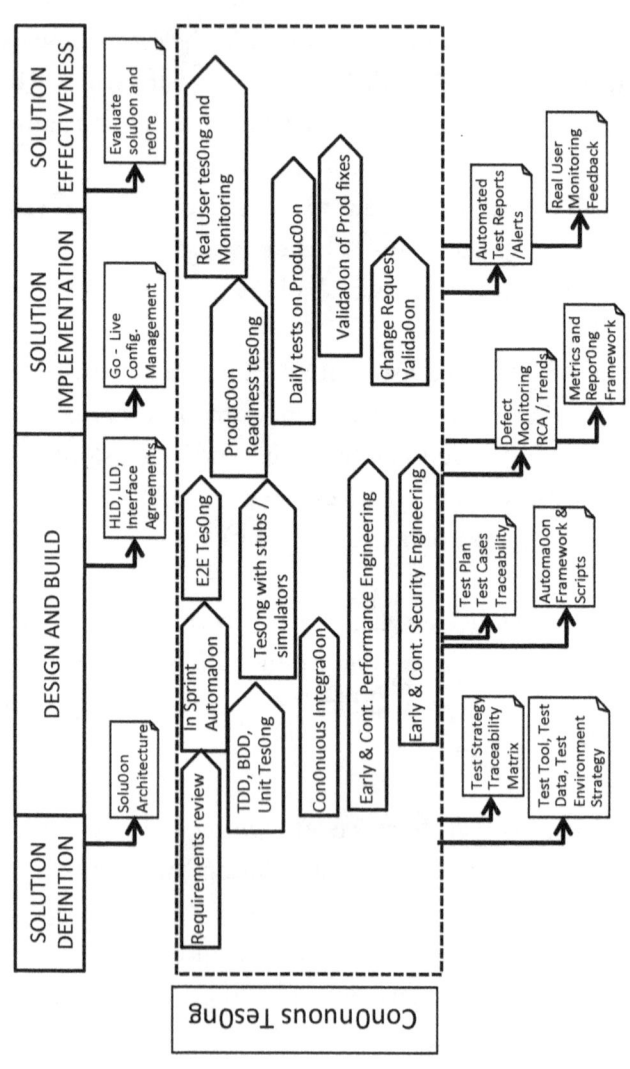

As Mark continued to describe how they were planning on elevating themselves as a Quality Engineering team, he fielded questions by his managers and leads on the Test Driven Development. He went back to the board and wrote what exactly they would be doing and corrected himself and everyone, he said Behavior Driven Development was the term and showed how they would be going about it.

"We adapted to the shift left into Agile process that we formed a couple of years ago by doing what precisely?" It was rhetorical than being a question, so he didn't wait for a response, he continued with " We write test cases and test steps when the user stories first come into the sprint, then we begin execution."

Now with Behavior Driven Development, we are going to shift further left. It is not as straight forward, but the process will be similar. Let me explain. When the business says I want new functionality, we break that piece down into user story, sometimes multiple stories. Then, we take each of those stories, convert them into features, and break those into scenarios and then into steps. These steps et al become your primary acceptance criterion for code development to occur or for you to get a sign-off from the business team.

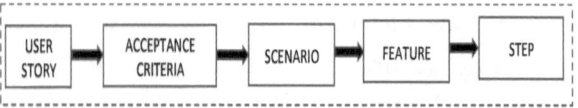

Once we have the sign-off from business, we provide these steps to the Dev team member we are assigned to, one at a time. The goal is to develop incrementally and validate each unit thus implemented.

The developer then builds a unit of code with an intention to make the test case pass. Now, the Dev team member you are working with implements the code until you are satisfied that the particular method or group of methods working together satisfies the requirements for a specific "Step", "Scenarios" and in turn the feature, which is the essence of what we are trying to achieve.

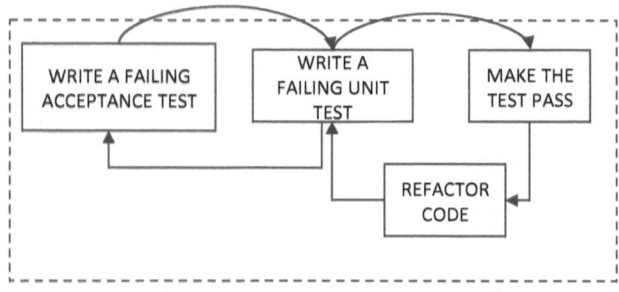

"The concept we will be implementing BDD, to put it bluntly, this is hybrid of Acceptance driven

Development and Test driven Development." "He used Dave's race car analogy to further clarify how he meant.

The Automation manager raised his hand with a puzzled expression. "This might work well for units of code, but wouldn't it be too hard to maintain and run tests at the time of build?" "Good point" Mark said and reiterated that the tests were English language like and explained how the gherkin way worked to his team. He further explained how these tests would be grouped as features and they execute as a suite of automated unit tests for groups of functionality. Also, he went on, now we can make use of these methods with a combination of page objects to build automated tests at functional level.

"So", he said, "we have automated tests for all units developed and mapped to stories / acceptance tests, and we will have automated functional tests that will be developed as part of the same sprint and readily available. All of these can be grouped and executed for every build. How can we automatically do it? This is something I am working with Tom and Phil on, we will talk more on that shortly." 10 minutes guys, I'll see you all back here. He walked over to meet Tom. He explained the plan briefly to Tom, who liked what he heard.

Tom pulled Phil and the trio headed to the conference room. As the team got back from their break, Mark briefly explained what they had come up with thus far, and what lay ahead. Phil who was responsible for configuration and release management listened intently.

Phil had joined the organization precisely to build something like this. He had done several of these implementations before. Phil said, "There are two things we can do when a build is ready. We can run all the tests your team created automatically and report the status at the end of it. This process is called Continuous Integration and it can be set up to trigger on a "build event" or it can be manually triggered. The best part is we can specify what environment you want them executed on, provided the tests are setup to scale that way." Phil took a pause and looked at the gang, he could see their thoughts churning away.

The excitement was building in the team; they eagerly shot questions at Phil, who was equally excited to share the knowledge.

Here is a simple process flow on how things work he explained.

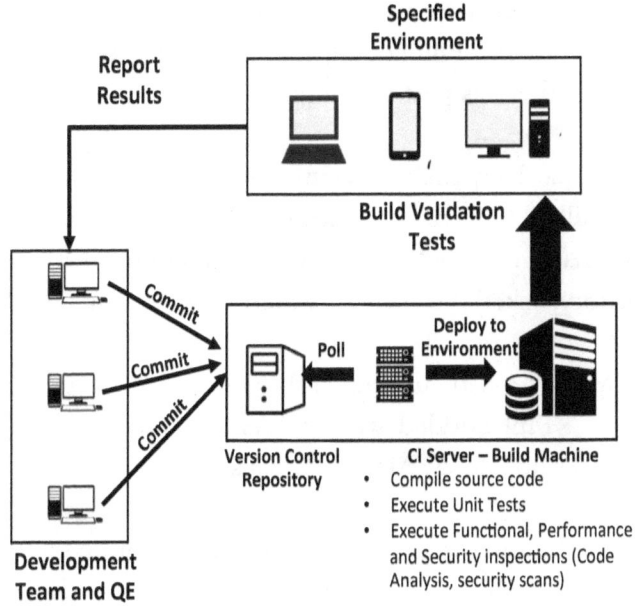

Ahmed who was responsible for performance testing asked, so Mark " Does performance testing still have to wait for pre prod?" "Good question" Mark said, Continuous Testing will include early performance testing which would essentially be 20-50 user load tests that would be used for Code/Query optimization." Phil added, "and the best part, we can include them as part of continuous integration job!"

"What happens to our number one challenge? The unavailability of environments?" Asked the automation lead. "And how about compliance and security, asked the security expert in the room Tom"

Mark looked at Phil and said, "Can we include those". Phil nodded with excitement and continued. "Oh yes absolutely Mark". "That's perfect"

Mark now turned to the group and continued, "So we can include functional, performance, security analysis techniques, static and dynamic. We can identify code level issues with the static analysis and vulnerabilities and penetration testing will be more of automated tests. We will need to get creative around how we handle compliance if you mean 508 kind of compliance. There may be a way to do it."

"Coming to the environments", Phil continued we can leverage some of the more modern techniques to automatically build and deploy environments with

Predefined configuration parameters and essentially deploy the build onto such environments and execute all tests within; all this can be performed as part of CI. But we will need significant help from Tom's boys. We can take that offline.

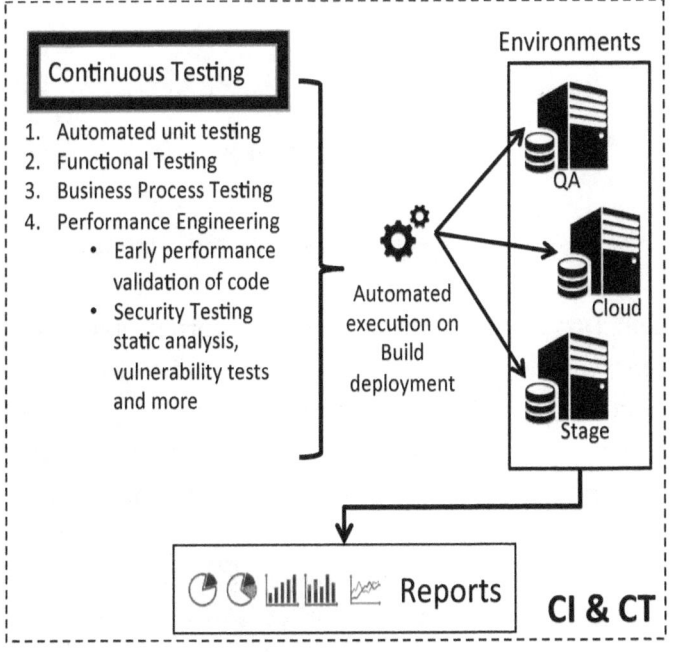

Continuous Planning, Continuous Development, Continuous Testing and Continuous Integration.

It was all coming together.

*　　*　　*

Authors note

This approach requires significant amount of coordination, collaboration and of course cross-skilled teams. One thing that is important and often times gets ignored is that it is imperative to have a team that gets the big picture and for the leadership to have a vision. This sort of approach will work if it is a top down push and a bottom up implementation.

So you – Yes you, with this book in hand, you play a key role. Believe me, it neither is as difficult as you think it is; nor as easy as I make it sound.

Imagine this, the first time you sing in a choir or perform in an orchestra, It seems extremely complicated to get in sync; Well some are naturals. [But we are going to assume we don't have all Mozart calibers on the team]. It takes practice and tremendous amounts of effort from the conductor (Leadership in our case) to keep everyone performing their roles in Orchestration, for the symphony to sound the way it does.

With quality as the end goal to drive and align with business goals, It is very important for every person on the team to know that testing as a chore or activity at the end of the lifecycle is dead and to ensure Quality at each stage in the life cycle is their responsibility!!

This whole change in game is what Quality Driven Development (QDD) is all about. In fact The Chief terms it as QDD + Continuous Deployment, Monitoring and Feedback as the core essence of Continuous Delivery.

Oh my bad, I am getting ahead of myself again!! ☺

I will let you get back to it!

* * *

10

Chief,

We are on Target with the strategy and a detailed plan for Mission Continuous Delivery. We have an all leadership meeting during which we will consolidate the methodology and have a true end-to-end strategy. More on this during our meeting.

Thanks,
Dave!

Dave read the email once more and verified his CC list before clicking the Send button. He leaned back in his chair and was going over the plan for the

6th time with himself. He was content knowing that he had a plan, more importantly a plan that would work. Being thorough as he was, he wondered if it were really airtight or if there were loose ends. The Navy seal in him never let him rest easy. It was always the unknown that fails the best of plans and he wanted to be sure he had covered all the bases. He sent a quick IM to Kumar to check on the status of the team make up for Continuous Delivery and as an afterthought added that he needed the skillset matrix that they needed to give Mark, to help Mark be ready with the team.

Across the floor, Tom was in his office, in an intense and animated conversation with his leads. "We need to set up Automated Deployments," his voice thundered in his office, as he continued to explain. Inspired by the rest of them, he changed his mantra to Continuous Deployment, Continuous Monitoring and Feedback. As for the agenda, he narrowed it down to 4 focus areas from the earlier 8.

1. Continuous Integration Environments (Lower and Cloud Enabled)
2. Release and Deployment Automation
3. Pre-Production Environment
4. Continuous Monitoring and Feedback

He turned away from the white board and continued in his deep voice. "So, Boys, I need each of you to own a focus area and make it the best there is. Technology is available to you and I want you to make Innovation the name of the game here. There is no such thing as a DevOps team – what we will do is adopt all the philosophies of DevOps and repurpose the members into each of the focus areas"

Lets start with Continuous Integration. We have the existing QA environment, we are going to have another environment that would be a replica as a back up until we have figured out a way to set up on-demand and automated cloud based environments. We will be requesting fresh budgets to revamp the pre-production environments, not necessarily to the scale of production but with similar infrastructure, configurations etc. to have the binaries ready for deployment. The policies on pre-production will be similar to production. Release and deployment automation will be limited and incremental until we learn to trust our process, environments and code – but most importantly until we have a foolproof way of doing things consistently. With no questions around the room, he continued.

Continuous monitoring is something we have implemented and in pockets. We do a good job of monitoring the servers, we've always looked at it as a NOC / SOC activity but now, I think we will have to get to a true Shift-right approach. Understand transactions, business processes, customer experience and more. We will work with Mark and team to set up scripts that can run in production, capture the details needed proactively and setting ourselves up with an ability to react to customer feedback quicker. This is Trend-Shift! We shift right to give the teams on the left more focus, control and visibility. With this, he concluded his passionate speech to the boys, they seems geared up and ready to jump.

Dave was still in his chair thinking out loud "The problem was always the unknown, specially the kind that you don't know, that you don't know". He heard a distinctive knock on his door that interrupted his thoughts.

"Knock, Knock, Knock"

It was the kind of sound made by metal on wood, not your standard wedding bands but thicker school rings. The chief was one of the 5 people in the building that wore their college school rings. The Chief's was an SMU Mustang Pride, his class of 1994 ring with SMU Dallas hall on it. It's got to be the chief, Dave thought.

"Come on in" he said.

"Hey Dave," the Chief said as he walked into Dave's office. "Got your mail. I was on the floor and thought I'd stop by. We are all set for the meeting,

Peggy is moving some stuff around on my calendar and you should see the invite shortly". The chief was not one for small talk and Dave wondered what it might be that brought him here.

"I noticed something and wanted to talk to you about it. I see some change in the way folks are moving around. The body language is stiff. We need to send a strong message from Technology Leadership that Change is Good! Remember, I can't emphasize enough on the fact that Culture and collaboration can make or break this whole effort that's underway. Let us plan on an IT all-hands once we are ready with the announcement and ensure we do sufficient brown bag sessions and such over the course of the quarter to get them all working together. That's all I had Dave, I'll let you get back to the grind". As he prepared to leave, they heard another knock on the door. It was Kumar. Kumar and The Chief shook hands as the Chief was leaving. The Chief thanked Kumar for all the wonderful work his team had put in and checked in on how things were going with him. He said, "I won't hold you guys up, If you need me, you know where I live" and walked away.

Kumar had come in with the Sprint plan, skillset matrix and the responsibility matrix. Dave reviewed from the printed copy and approved the plan with a couple minor edits.

Mark met with his team in his office to review the skillset matrix for both his FTEs and the vendor team. He suggested the vendor to get back with a revised team make-up and proposed onsite and offshore team structure. Everything looked perfect and he had a sit down with his trusted lieutenant on the tools needed to get started with Continuous Development and Continuous Testing. Before they started discussing the tools, they had one important business to discuss. Current automation.

One of the key issues he had had with the automation currently underway was the extreme dependency on the UI. The team explained to him that the current automation scripts would run only on one environment and that it was significant effort to make them portable or even for them to run it on multiple browsers in parallel. On digging deeper, they identified the issue to be in the way the scripts were developed, they were all "Record and Play back" with hardcoded values for browser, environment and in some cases even test data. The concept of a framework was alien to this bunch! He instructed his Lieutenant to come back with a comprehensive plan for automation this time around. Especially with BDD and Early automation and all the fun stuff he was planning for them to do, having the right skillset, toolset and framework was important.

His team was already working on building a proof of concept using some of the new age tools for BDD, automation, continuous integration plug-ins and more.

Mark spent a few minutes looking at the demo of the POC, which was built on a sample open source application for hotel management, which was taken from Github. The demo impressed Mark greatly and he agreed on the shortlisted tools. He told his team that they would next showcase the demo to a few folks from either side of the value chain, meaning Development and IT Operations before finalizing the tool from the shortlist.

Satisfied with the progress made thus far and feeling confident about the skillset matrix that his team had come up with, Mark headed out to the meeting.

He met Dave and Tom in the pantry as they all grabbed coffee and off to the 12th floor for the leadership meeting.

Each of them discussed their plans and this meeting was only to summarize what they intended to work on. It went without incident, just like old times, a little humor, an argument here, agreement there and progress all around.

Author's Note

Oh Hello Again. Dave is right; he should be worried about the "I don't know, what I don't know" part. Case in point, you can plan for what you know, be prepared to tackle the stuff that you know that you don't know; but how on earth would you be prepared for the stuff you don't even know that you don't know!

Heck, even Achilles was only as strong as his heel!

In my opinion, 2 things should worry these folks.

1. Quality Intelligence
2. Too many tools

Metrics is one side of Quality Intelligence, which needs to be talked about. You cannot measure what you don't track, nor can you analyze what you don't measure! It is of tremendous value that this sort of analysis can bring. Metrics don't lie! They bring out the real story.

With such an endeavor that they have undertaken, I think the biggest benefit would be from correlating the different stages within the lifecycle and metrics from those to gain additional insights. This could be their opportunity to understand the Cost of Quality, Cost per defect and even help them understand the ROI of the initiative and predict future releases – the whole nine yards really!

It is much bigger than selecting the right toolset. What an Enterprise really needs if they aspire to get to Continuous Delivery is much beyond a Toolset. It is like Orchestration – You need the tools but more importantly, an Orchestration engine that can control / manage the tools.

* * *

11

Over the next few weeks, several fruitful discussions emerged as they all worked towards the finer details for the end-to-end methodology. The process was laid out and accepted mostly, except the chief hadn't seen it yet. Team and sprint structures were being revised and the teams were piloting some areas where the teams had to work together on a varied set of tools, to conduct proof of concepts on the tools that would bring the most value and be the most disruptive.

The technology leadership met for a daily stand-up these days and sometimes had extended sessions almost everyday this week. The theme of each of these very healthy discussions was similar. They discussed budgetary impacts, seating arrangement, offshore arrangement, coordination and awareness programs to get wider buy in and success.

The technology floor was buzzing with activity and excitement this morning. They all knew today was the big meeting. The Technology leadership was all huddled up in Dave's office all morning.

As they approached the hour of the meeting, all of the Technology Leadership was still in Dave's office going over the presentation deck they had put together for that afternoon one final time.

As they reviewed the Continuous Integration plan, Tom said, "This is going to be a game changer right here and its next only to user experience monitor".

"Oh Yes" Mark said sharing the excitement.

Dave said, "This is great" as they approached the end of the presentation. "This change is not going to be easy and it won't be perfect, it probably will be a sprint or two before the bruises our teams cause to each other heal and they start working well together" "and perhaps even longer for all of this to sound like a symphony", added Tom.

They all laughed as they headed up to the 12th floor a few minutes before the hour. Each of them excited and thinking about what the future held.

As they got into the elevator, one thing still lingered in everyone's mind, what have we not thought about? Well, budgets weren't seriously dented, team layout and plan wasn't as bad, everyone agreed that they perhaps had too many tools in the mix, Phil suggested training was important even though they were all techie folks who could pick things up and run and ensured they added a training plan and budget, environments were going to be a changer.

Still no metrics

No thoughts on consolidated views from various tools

#justsaying
#Orchestration

*   *   *

12

The twelfth floor conference room smelled like sweet barbeque sauce and Mark smiled at Dave. The Chief sure knew how to keep his "A Team" happy, Good old Texas style bbq for lunch!

Peggy was watching over the catering team as they set up lunch and the technology leadership walked in. "Just the best for y'all" said Peggy with her exaggerated southern drawl. She informed them that the Chief would be a few minutes behind schedule and that they should help themselves, not that they needed an invitation. As they all got their plates full and settled in the plush chairs of the boardroom, the Chief walked in, "Howdy Fellas" he said. "Good stuff Chief" Tom said, to which the Chief replied, "Oh Thank Peggy, she picked the spot and the menu".

While the Chief filled his plate he asked, "What do you make of my Cowboys' chances this year" asked the chief referring to the Dallas Cowboys. "Oh absolutely great chance, this will be their best year in a while" replied Dave enthusiastically. Everyone shared the excitement except Tom who was a Niner (49er) for life and Phil who had just moved from Patriot nation. They joked around as they finished their lunch. Peggy stopped by to get the food trays removed, refilled a freshly brewed pot of coffee. They grabbed cookies and coffee and settled back into their positions.

They all looked to Dave to kick things off. Dave began, "So, we hung on to the theme of Continuous in defining the process framework, essentially it will be Continuous Development, Continuous Testing, Continuous Integration, Continuous Deployment, Continuous Feedback and Monitoring". The Chief smiled approvingly.

There are a few fundamental changes we need to bring to the way we are all aligned, Dave began. First, I think we need to do away with our definition of lifecycle. It no longer can be Software Development Lifecycle – I think it brings a siloed thought process. We will call this our Software Engineering Lifecycle. Second, We will do away with phases it makes it sound too waterfall like. We will accomplish our tasks within the SELC in stages. We will start with our

Mantra as an enterprise, well one of the few we will have – Quality is a Responsibility and we are all stakeholders.

To complete the mission Continuous Delivery, the goal we are going to start with is to have a deployment ready build at any give point during the cycle, this we think is paramount to our success.
The chief asked a few pointed and valid questions and he suggested that they essentially were getting into Quality Driven Development as their way of getting builds ready followed by continuous deployment et al. Dave smiled at the simplicity with which things were summarized.

Next, Dave started to dig into the methodology. He said, for starters, this would be extremely collaborative development approach. Two primary benefits from it. One, Accelerated Development with higher quality and Two, Reduced and in some cases eliminated wait times. Let me explain how, he continued. We will begin application development based on Behavior Driven Development as the philosophy and with one simple premise. All code is guilty until proven innocent. How we do it is by bringing the erstwhile test team, now called the quality engineering team because of the revamped team structure will have pointed inputs for each piece, unit or method of code being developed and in two parts. One when they give the acceptance criterion based on user stories and two, when they test and validate those individual units developed. This will be continual process and incremental through out stage. All the units developed and validated are added to the check-in repository, Mark's automation unit will focus on building in-sprint automated test scripts. They will leverage the methods developed to ensure comprehensive coverage and accelerated automation.

Additionally, the non-functional testing crew is building incremental load tests, security tests and other non-functional tests for each release. All this ensures that we have 3 types of tests by the time a build is ready. One, Automated Unit Tests, Two, Automated Functional Tests and Three, Non Functional Tests that focus on specific functionality, code and query level issues and native platform issues from a security stand-point, all these ensure we don't have heartburns later in the lifecycle. The best part Chief is that all these tests are automatically triggered and executed on approved check-in and may be reused down the line when we get into regression mode. All this can be set up through the Continuous Integration server that will be set up in collaboration with Tom's team. So at the end, each passed build or a series of passed builds can be moved to pre-prod, further tests run and we can essentially have a production deployment ready binary!

The chief nodded to everything being said thus far and frowned at the mention of pre-production environment. Tom who saw the reaction immediately chimed in with, you are right, we don't have the environment the way we need it to be able to do what we intend to. But that is something we had to tackle anyways, so better now than ever before. The reason we are calling for a stronger pre-production environment that can be restricted and left pristine is to enable stringent configuration management practices and further carry out production grade tests in this for hot fixes and such.

All this chief is coming with little to no addition of head count, said Mark. We are working with the current vendors to restructure contracts where they can fulfill and replace with engineering vendors with adequate knowledge transfer where they are unable. The quality engineering team going forward would be composed of SDET's and Quality Engineers and a limited few manual testers. Further more we will support Tom and team in building the continuous monitoring and feedback scripts for swifter reaction times.

"How does all this sound?" asked Tom. We have revised sprint teams, plans and once we have a formal go, we will work on rearranging the seating maps, have the all-hands by you, conduct awareness sessions for the teams for a couple weeks and kick things off later that month. The chief nodded his approval for the most part and had questions that were fielded by whoever were best equipped to address.

I Love this, great work guys, the chief said. One last comment I have that you may already have thought about, we would be introducing many new tools I imagine. Yes, said they all in unison. How do we ensure they all talk to each other and we always have one source of truth? How do we ensure we have control and visibility across all actors / stakeholders seamlessly? Who is the Conductor to our Orchestra? Dave smiled and said he would work on that. Metrics was another open item that the chief has asked the team to get back with, Mark volunteered.

They had the approvals they needed; now it was go time!

* * *

13

Everyone had gathered in the town hall and as they waited for The Chief, Dave and Mark discussed the toolset that their "prototype team" had done a proof of concept for. Tom and Phil were talking about some of the findings of using AWS as their cloud for on-demand environments and some of the challenges they had encountered.

The Chief walked in at the top of the hour and began with "You are all probably aware why we are gathered here today, or are you still wondering?" he joked.

We have chosen Technology because we chose to do the impossible; we as a technology team here are equipped with the resources to do just that.

Innovate or perish, trust me when I say this, We are never stagnant, we are either growing and maturing or are falling behind. I encourage each of you to try and simplify the way you do things, that doesn't mean shortcuts but to take an approach that will optimize your use of time.

Remember to learn new technologies and to keep honing your skills, we are all like woodcutters, we can either take a pause to sharpen the axe or struggle with the axe of knowledge we have, which becomes blunt over time. So, take a pause to learn a course, share the knowledge and enlighten others as well.

Adhere to the core philosophies of engineering and make Quality a habit. Just like Frodo was a creature of Hobbit, we in technology are creatures of habit, form them and form them right.

That is my list so far, of all those one in particular stands out to me on why I chose technology. I want to build the next generation applications that would help us succeed as an organization. In fact, I would like to think that is why we are all here. We have a platform, a solid foundation, the resources and most importantly, the passion and energy to do just that within our organization.

So, in the words of Sir Isaac Newton, If I have

seen further it is by standing on shoulders of giants, as a retail giant we are proud of the several firsts in the industry but the one I find particularly exciting is our culture, our tradition of innovation and excellence. It is not our job to follow in the footsteps of what we have done before or on prior releases but to Exceed it, crush the shoulders of the giants upon whom we stand and go further." When he was done, he got nothing less than a standing ovation.

With that impassioned speech, The Chief had set the stage for the rest of the session. He then summarized what the goals set were and how to get about to achieving them. Mission Continuous Delivery, he explained, was aimed at improved agility and sustained velocity to assist the internal teams to meet and exceed the business goals.

Align Release, Quality and Business Goals

The Chief explained, "The idea is to align the way you plan, build, release and deploy applications. Two things to consider here; A trend shift™ approach is necessary; focus has to remain on the right (production and release) with a continuous feedback loop to planning and business phases on the left. Additionally, he pointed out.

1. Focus on quick wins
2. Insist and make incremental changes

3. Emphasize on continuous improvement
4. Let us not go big bang when it comes to changes
5. Avoid massive rollouts

Define Critical Success Factors

To begin with, understand what the business wants? What was committed to or by the marketing teams? Gain visibility into these areas and ensure you have a tracking mechanism for the metrics and you have the needed horsepower to stick to the plan and minimize the gap between plan and actual.

1. Align the success factors across the various players in the value chain
2. Architecture visuals for signoff by everyone involved
3. Collaboration and clearly assigned roles and responsibilities
4. Impact analysis

Automate, Automate, Automate!

According to my formula for continuous delivery – The manual activities in the lifecycle are inversely proportional to the velocity of continuous delivery. In other words, let us minimize our manual activities, eliminate wait times, collaborate and ensure smooth hand-off and communication between teams. This will be our mantra:

1. Continuous Development – Test Driven /

Acceptance Test Driven Development – Plan, Test, Develop, Repeat
2. Continuous Testing – Begin with Automated Unit Testing, Early Automated Functional Testing, Early Performance Engineering and Security Engineering
3. Continuous Integration – Build automation, Continuous Integration and deployments on Pre-Prod environments
4. Continuous Deployment – Have prod like environments that can be created on demand, where all of the tests can be run. Build configuration and environment parameterization at the start of the cycle (plan)
5. Continuous Feedback and Monitoring – Understand application production behavior, enables teams to be proactive by building an automated feedback loop into plan phase

Scaled Pre-Production Environments

One of our biggest challenges has been an inadequate pre-production environment, Not Any More! A stronger pre-prod environment that can scale and be utilized by all teams in the value chain is much needed and Tom's team is going to work on making it better.

1. Integrated pre-prod environments
2. Integrated and collaborative engineering and

ops teams (TestDevSupOps)
3. Automated and on-demand environments with ability to auto deploy
4. Efficient and effective tool utilization and cross- skilled teams
5. Enforce release repositories with version control (Store builds, Binaries for quick rollback)

Monitoring, Metrics and More

End to end lifecycle metrics are to be tracked, measured, analyzed and used as continuous improvement feedback. Production monitoring, real user experience measurement will become the norm. We will gain more insights into application performance in production, to help us be proactive in understanding the feedback.

* * *

EPILOGUE

As the steering committee geared towards the first full week of implementation, one marked difference in the way sprint planning occurred was evident. The Chief along with all the technology leaders and their respective Architects and Tech team Leads were a part of the discussion. The planning session was held over a four-hour session and at the end of it everyone agreed that it was one of the more thorough plans they'd had in a long time. Each of them had come prepared with their agenda, the ground rules had been laid by The Chief. It was a matter of implementation and the team did all they could to ensure a seamless transition.

This enterprise was no exception; it took them 2 sprints to get the concepts of BDD and a release to get the improvements in the process trickle down. After a couple releases, things settled and this process became a well-oiled engine. The critical success factor for this as with anything was top down initiative built bottom up with executive commitment, governance and most importantly, wide acceptance by the teams. They went through their fair share of attrition on the team, specially in the transition from a Quality Assurance team into a Quality Engineering team, but what stood out is the perseverance with which they hung on to their mission and made it happen. This enterprise now had the ability to deliver on demand, a fully automated test suite; they monitor real user experience and much more.

Should they be content? Should they stop here? Well, it would be a case of technology stopping at this.

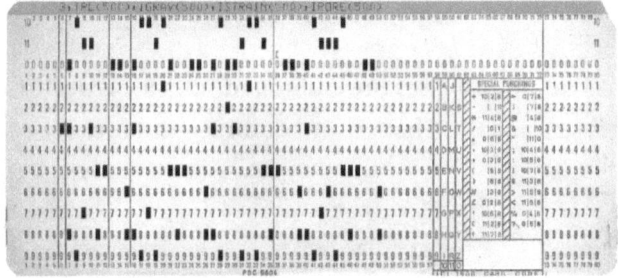

For those of you that do not know what this means, these are punch cards, Wikipedia describes them as

"A punched card or punch card is a piece of stiff paper that contains digital information represented by the presence or absence of holes in predefined positions. The information might be data for data processing applications or, in earlier examples, used to directly controlling automated machinery in the 20th Century". Vow! 20th Century they said, really? That stung a bit.

The next generation holds a lot of excitement, Artificial Intelligence for automated problem solving for starters. Artificial Intelligence will have a huge role to play in the way Application Engineering Teams (Dev, Test, Sup, Ops) work in the future. Automated problem solving will be applied even to areas such as code level issues, automated environment management and more. So, you either innovate, adapt or perish.

The possibilities are unlimited, it is a matter of our ability to adapt, change and embrace the future.

* * *

APPENDIX I – METRICS

Metrics is a very delicate subject at most enterprises I have worked with and within. It is often thought of as a fact-finding or worse faultfinding exercise by the teams that report the data. But then, if you really think about it, How would you measure something you do not track and how do you analyze something you don't even measure and most importantly how do you improve if you don't begin to understand what went right or wrong?

Quality Intelligence and Metrics in the Application Engineering Lifecycle are the one way to keep everyone in the value chain on their toes and assist in steering course as and when needed. Metrics specially the kind that are predictable can enable IT leaders and application teams in better planning and enable critical decision making, proactively. Just as an athlete or a sports team needs to understand the quality of the effort, statistics to continually improve an enterprise needs adequate means to track, measure and analyze metrics for continuous improvement.

This is what Quality Intelligence is all about! In an era where quality has taken center stage in how your customers view your brand, it is more important than ever that you ask your teams to track the data relevant to the 5 metrics being discussed here. Note that these metrics are not listed in any particular order, but when the information from all these is correlated it would paint the IT leadership a perfect picture on where things stand.

1. Productivity Metrics

Productivity metrics are crucial in more than one way. One, to understand where you stand in your current design / execution cycle, and two to accurately estimate the effort involved for future cycles. Not all applications or projects are born equal, so one formula at effort estimation is rather ineffective; thus arises the need to accurately capture the burn rates, design and execution productivity and environment parameters that influence the ability of the teams to be productive! The appetite of variance varies by the enterprise and hence the thresholds must be appropriately assigned.

2. Efficiency Metrics

Enterprises often mix up efficiency to productivity – there is a fine line between the two or a significant one – depends on how you look at it. Point being, they need to be measured differently. Efficiency grows when a task is performed repeatedly – for instance, the more often a test suite gets executed, the faster the team can complete it – technically at least! One of the key things to consider when working on improving efficiency is Automation. Consider measuring Design and Execution Velocity, Automation Efforts Vs Manual effort reduced, Defect detection efficiency, Release and deployment efficiency, environment parameters, down times, planned Vs Actual metrics for design and execution etc.

3. Effectiveness Metrics

You testing team could be spending hours in test execution producing near perfect results for the QA environments, but then – none of those modules they are executing tests on were touched; or on the other hand, they've had too little time to test and resulted in delayed release or worse yet – production defects. The point is not to eliminate regression tests entirely, but the goal is to optimize the effectiveness of the test efforts.

The effectiveness measurement is important in understanding functional vs. regression effort, defect yield, defect density, impact analysis, coverage analytics (Requirement traceability, Code coverage, functional coverage), performance metrics, security analytics and environment metrics (It is critical to ensure that the tests would pass on any environment – Literally).

4. Defect Metrics

Perhaps the most important of all! Capturing defects alone is not enough. Classification of defects based on their detection in the lifecycle is significantly of more value than knowing that a defect showed up. Defect metrics must include defect analysis, defect severity, defect age, when the defect showed up originally, defect rejection rate (represents quality of testing team), defect RCA, defect distribution by module or similar, defect status, defect reported by automation Vs. Manual test efforts, regression vs. functional etc. Additionally, it is imperative to understand and report the cause of defect such as environment, data, requirement and much more. This list could potentially be overwhelming, but it is the source of truth.

5. Production Metrics

Most times the best way of understanding your

application quality is your production metrics; it often becomes the scorecard of not just your IT organization but also your brand itself! The production metrics that must be considered are performance capacity planned Vs Actual run time, response times, real user experience, synthetic monitoring, uptime, defects in production proactively detected, defects reported by customer etc.

One of the significant factors is the ability to accurately report the production defects. Case in point, at a large enterprise that we interviewed and subsequently implemented metrics for, the QA / Testing vendor consistently reported zero production defects for 2 years running. However, when we went down to the source of truth – their production and issue reporting system, we noticed about 10 critical defects that went unreported. The argument from the vendor was that "We knew those defects would surface". This thought process is fundamentally wrong! You must ensure that all production defects are captured and reported for analysis; this could be the first step towards continuous improvement if analyzed accurately.

* * *

ABOUT THE AUTHOR

Prasanna Singaraju is the Co-Founder and Chief Technology Officer at Qentelli, A Dallas based Continuous Delivery Company. He brings vast experience in Quality Engineering practices, Agile Transformation, Building DevOps practices and Continuous Delivery.

Prasanna holds a Masters degree in Engineering from Southern Methodist University in Dallas, TX and lives in the DFW area.

NOTES

Continuous Delivery

NOTES

NOTES

www.ingramcontent.com/pod-product-compliance
Lightning Source LLC
Chambersburg PA
CBHW021409170526
45164CB00002B/567